HOW'S THAT HUMAN?

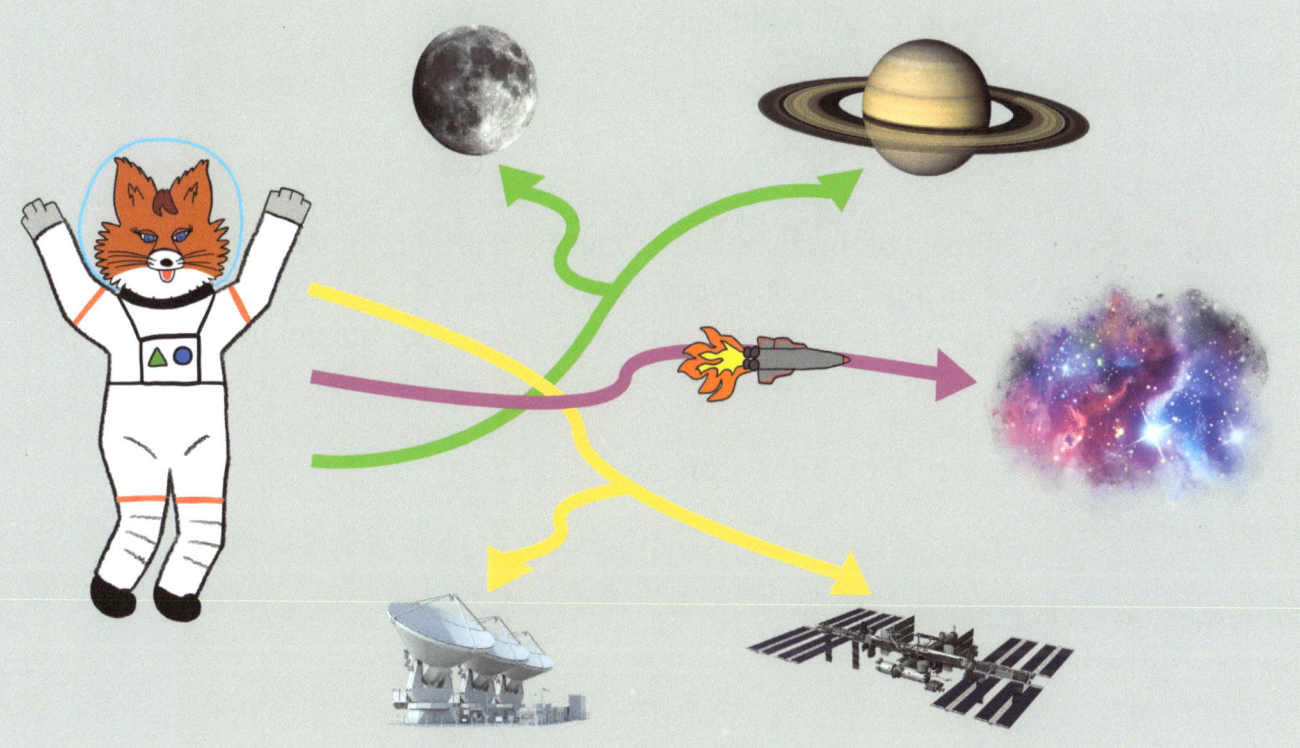

SPACE
A CHOOSE YOUR OWN EXPLORATION BOOK

HOW'S THAT HUMAN?

Cover Design: Rita Claire and AM Conroy
Opening Page: Rita Claire and AM Conroy
Illustrations: AM Conroy

Copyright © 2021 HTH Publishing Inc.

All rights reserved. No part of this publication may be reproduced, stored in a retrieval system, or transmitted, in any form or by any means, electronic, mechanical, photocopying, recording, or otherwise, without prior written permission from the publisher. No part of this book may be reproduced in any manner whatsoever without written permission.

How's That Human Space - 1st Edition (softcover)
ISBN: 9798462475702

Published by HTH Publishing Inc.
www.howsthathuman.com

THANK YOU!

Mom and Dad
I want to thank you for always believing in me and supporting my goals in life no matter what they were or where they took me! I love you!

Friends and Family
Thank you for your continued enthusiasm for my ideas! Thank you for not only pushing me but helping me grow, test, and develop this book!

My Team
I couldn't have asked for a better team to make this dream a reality! Your hard work was infectious; teamwork makes the dream work!

Former Students
I can't thank you enough honestly! You pushed me to be better and I am forever grateful for your influence in the teacher I am today!

YOU!
Thank you for taking a chance on a science book that looks at the human side of things and why science is ... well... YOU!

Introduction

Investigating space has applications here on earth towards medicine, industry, communication, and other technological advancements. Additionally, to even explore space we must understand what it does to the human body, why that happens, and what we can do about it. Pick your own learning journey to explore these concepts and more!

Do you want to travel throughout the solar system? Go to page 2...

Do you want to visit stars and galaxies? Go to page 11...

Do you want to learn more about space exploration and its human ties? Go to page 17...

Would you like to experiment from home? Go to page 33...

The Solar System

Our solar system has so many unique and amazing things! Not only do we have two types of planets, rocky versus gas, but we also have other objects like comets and asteroids. We have multiple probes, rovers, satellites, and telescopes getting closer to several of these and sending back data that is changing what we originally thought. We are even trying to get humans to live on Mars and the moon! How cool is that?! Our relationship with the sun and moon explains so much about what we see here on earth.

Do you want to know more about the earth-sun-moon relationship? Go to page 3...

Would you like to know more about asteroids and comets? Go to page 9...

Do you want to visit the inner rocky planets? Go to page 7...

Do you want to visit the outer gas giants? Go to page 8...

The diagram above is not drawn to scale. The distances apart and sizes are not exact.

Want to know how to model the sizes and distances? Go to page 33...

Data Dive

What is the relationship between the average speed of a planet and its distance from the sun?

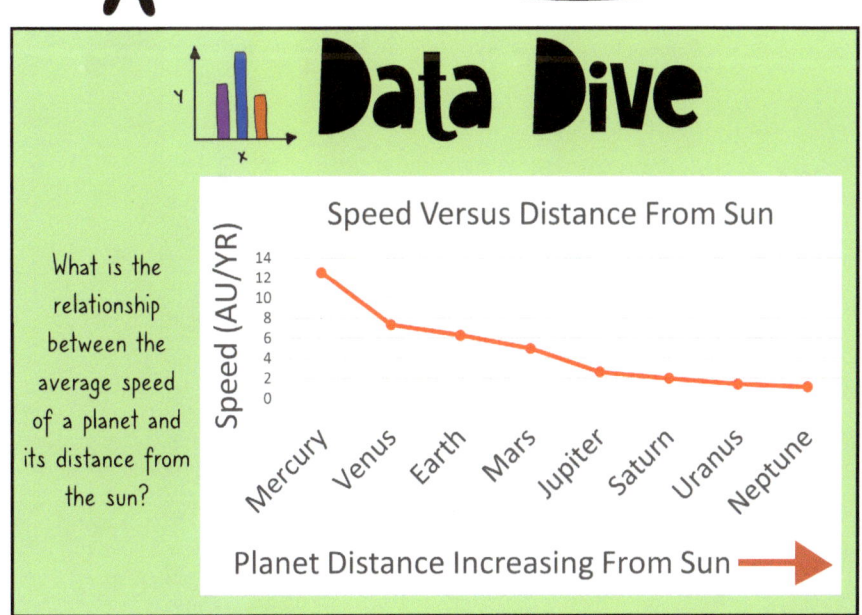

Speed Versus Distance From Sun

Planet Distance Increasing From Sun →

Scientist Spotlight

Nicolaus Copernicus was a Polish astronomer who was the first to come up with the model that the sun was the center of the solar system and not Earth.

The Earth-Moon-Sun Relationship

The earth rotates on an axis, spinning super fast. It takes 24 hours for a complete rotation. It may look like the sun, moon, and stars are moving throughout the day and night, but it is instead the earth that is rotating.

QUICK QUESTION: What is an axis?

-> an imaginary line around which an object rotates.

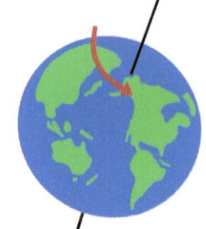

Would you like to model this from home? Go to page 34...

The earth travels in an eclipse shape around the sun for 365 days a year. As it makes this yearly orbit, we see many different things happening on earth in different areas.

QUICK QUESTION: What is an orbit?

-> a curved path an object takes around a larger object due to the attractive force of gravity.

Join the debate on the HTH Space book website! Which would be worse for the earth, losing the moon or a solar flare?

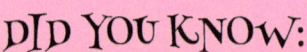

Do you want to learn about tides and eclipses? Go to page 5...

DID YOU KNOW:
The moon orbits the earth every 27 days and it also rotates on its own axis every 27 days!? Because they match up the face of the moon you see doesn't ever seem to change.

Do you want to know more about moon phases? Go to page 4...

Ready to Research?!
How did the moon form?

Search Suggestion: how was the moon formed

Do you want to investigate why seasons and weather occur? Go to page 6...

Moon Phases

The whole sunlit side of the moon faces earth in a full moon.

First Quarter

Through "waxing" more and more of the moon's surface is revealed.

Waxing Gibbous

Waxing Crescent

Would you like to model this from home? Go to page 34...

Full Moon

New Moon

With a new moon the sunlight hits the side of the moon facing away from earth.

Wanning Gibbous

Third Quarter

Waning Crescent

"Quarter" means you can see half of the moon's surface lit up.

Through "wanning" more and more of the moon's surface disappears.

Would you like to go back to the earth-moon-sun relationship page? Go to page 3...

Would you like to go back to the solar system home page? Go to page 2...

CREATIVE CORNER

What would you need for a colony on the moon? How could humans survive? Design your very own moon colony whether 2D or 3D and share at submissions@howsthathuman.com to be featured on the HTH Space book website!

4

Tides and Eclipses

Gravity is a very important force when we study the universe. Check out the HTH Physics book for more information on forces. A force is a push or a pull, and gravity plays a critical part in the behavior of our solar system, the galaxy, and other objects in the universe. Turns out objects naturally have a pull for each other, so distance apart matters just as much as the motion or speed of each object and how much mass or matter they have. Let's see how the pull of the sun and moon affect the earth.

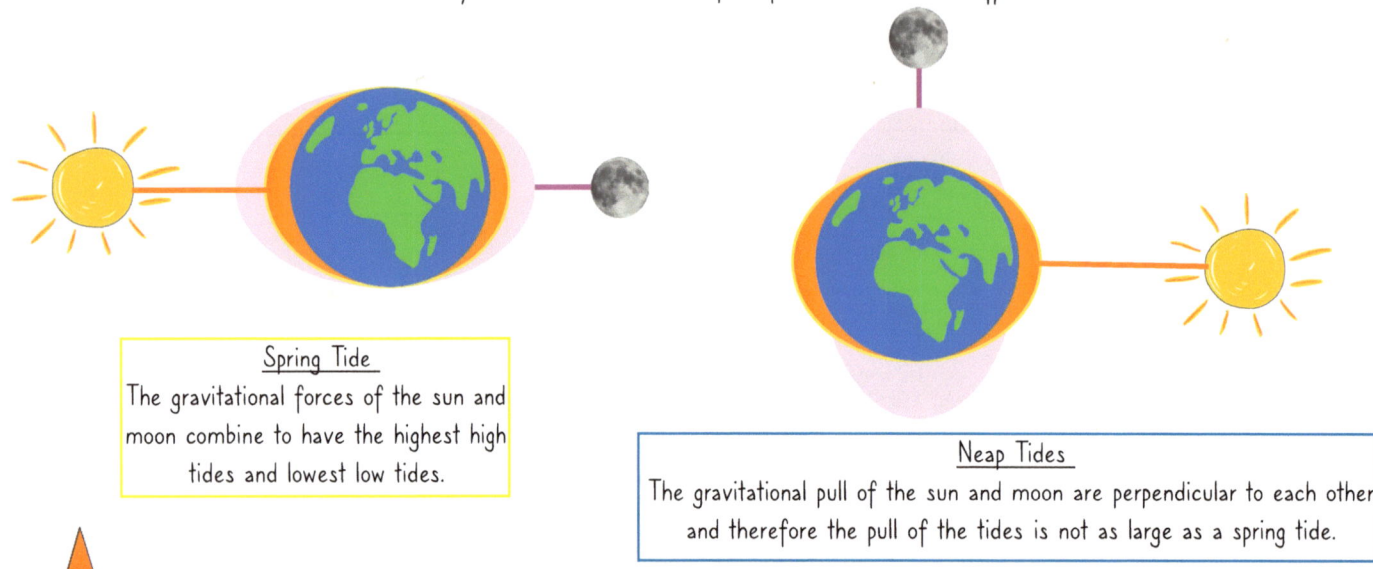

Spring Tide
The gravitational forces of the sun and moon combine to have the highest high tides and lowest low tides.

Neap Tides
The gravitational pull of the sun and moon are perpendicular to each other and therefore the pull of the tides is not as large as a spring tide.

QUICK QUESTION: What is Newton's Law of Universal Gravitation?

-> every object in the universe is attracted to other objects in the universe.

When objects are traveling at the right speed and the pull of gravity is just enough, orbit is achieved.
With the moon's orbit around the earth and the earth's orbit around the sun, we see some interesting things happen when the orbits overlap.

Would you like to model this from home?   Go to page 34...

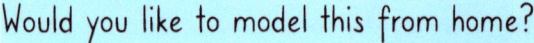

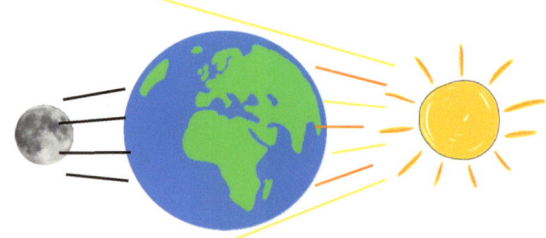

We have to be very careful and wear special glasses when looking at eclipses, why is this? Check out radiation on Pg. 28...

Solar Eclipse
The moon blocks light from hitting the earth.

Lunar Eclipse
The moon is caught in earth's shadow.

Would you like to go back to the solar system home page?   Go to page 2...

Seasons and Weather

The seasons are due to the fact that the earth's sphere shape lets light hit the earth at different angles, rates, or intensities. Direct rays of light hit the earth at the equator, elevating temperatures. All that energy from the sun's radiation fuels the weather and climate patterns we see today. The tilt of the axis also plays an important role, causing the continents to each face the sun directly at some point during the year long revolution around the sun. Tilted towards the sun, the season will be summer. Tilted farthest away, the season will be winter.

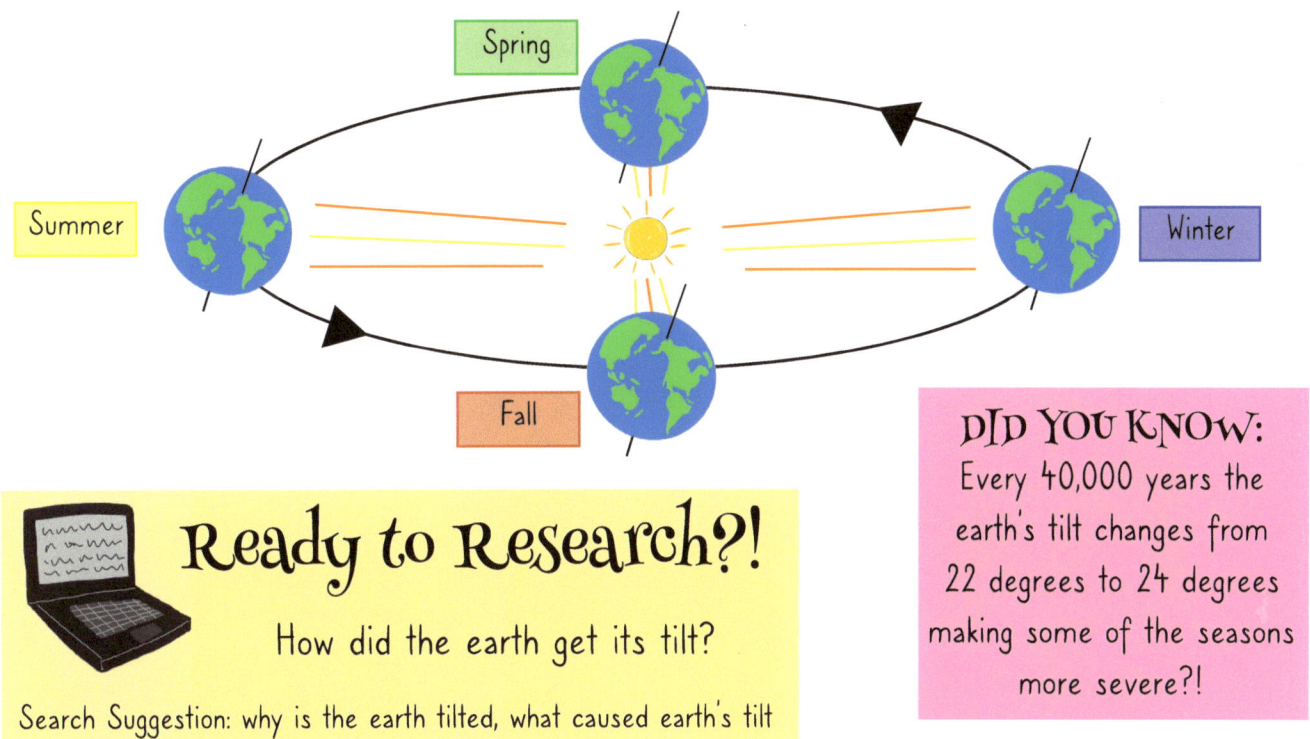

Ready to Research?!

How did the earth get its tilt?

Search Suggestion: why is the earth tilted, what caused earth's tilt

DID YOU KNOW:
Every 40,000 years the earth's tilt changes from 22 degrees to 24 degrees making some of the seasons more severe?!

Data Dive

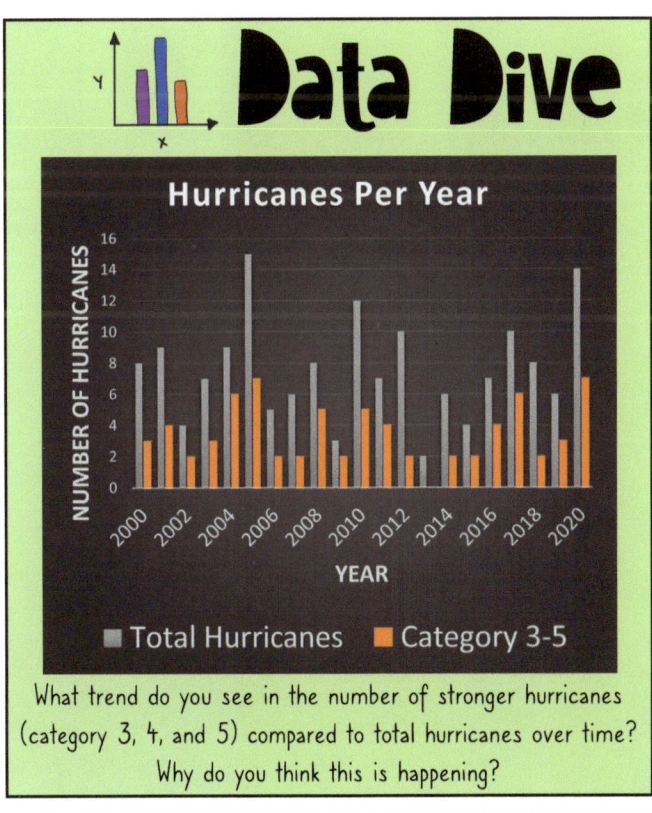

What trend do you see in the number of stronger hurricanes (category 3, 4, and 5) compared to total hurricanes over time? Why do you think this is happening?

QUICK QUESTION:
What is weather?

-> the uneven heating of the earth's air and water, causing wind and ocean currents to move transferring heat and energy through the water cycle.

Would you like to go back to the earth-moon-sun relationship page?   Go to page 3...

Inner Planets

Investigated by the probe Messenger!

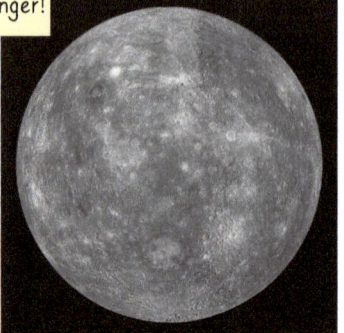

Investigated by the probe Magellan!

Mercury
1. No moons or rings
2. Smallest planet
3. A year takes 88 earth days
4. You weigh 38% of your weight on earth
5. The second densest and hottest planet

Venus
1. No moon or rings
2. Almost the same size as earth
3. A year takes 225 earth days
4. The second brightest object in the night sky
5. Hottest planet

Join the discussion! If you were going to Mars and only had a box that was 3 feet wide by 3 feet tall, what would you take? Let us know on the HTH Space Book website!

Check out the rovers investigating Mars, Go to page 22...

Earth
1. Densest planet
2. Only one natural satellite (moon)
3. Has a powerful magnetic field
4. 97% of the water is saltwater
5. Mainly made of iron, silicon, and oxygen

Mars
1. A year takes 687 days
2. Dust storms can last for months
3. The sun looks half the size it does from earth
4. Home to the tallest mountain in the solar system
5. Has 2 moons, Phobos and Deimos

QUICK QUESTION: What is a space probe?

-> a spacecraft with no crew designed to collect scientific data for specific missions.

Antenna technology for communication.

Solar panels to generate electricity.

Scientific instruments to gather data.

Probes have gathered data on all planets and several moons, asteroids, and comets. The Pioneer and Voyager missions along with New Horizons are set to gather data beyond our solar system.

Would you like to go back to the solar system home page?   Go to page 2...
Would you like to learn more about the technology used to gather the data?   Go to page 18...

Outer Planets

Investigated by the probes Juno and Galileo!

Investigated by the probe Cassini!

Jupiter
1. Largest planet
2. 4,333 days in one earth year
3. Has 79 moons
4. One day is only 9 hours and 55 minutes
5. The Giant Red Spot is a huge storm

Saturn
1. Orbits the sun every 29.4 earth years
2. Made mostly of hydrogen
3. Has 150 moons and smaller moon-like objects
4. One day is only 10 hours and 34 minutes
5. The rings are made of ice and dust

Investigated by the probe Voyager 2!

Uranus
1. Hits the coldest temperatures
2. Takes 84 earth years to orbit the sun
3. A day is 17 hours and 14 minutes
4. Rotates on its side
5. Has 2 sets of rings

Neptune
1. Has 14 moons
2. Atmosphere is hydrogen, helium, and methane gas
3. Most distant planet from the sun
4. A year lasts 165 earth years
5. A day is 16 hours

Investigated by the probe New Horizons!

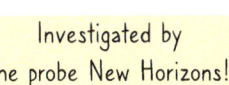

Dwarf Planets - Pluto
These planets are generally round and orbit the sun, but haven't officially cleared the path of their orbit like the large planets.

Ready to Research?!
What is special about the moons in the solar system?

Search Suggestion: solar system moons, exploring planetary moons

Would you like to start over?   Go back to page 1...

Asteroids and Comets

QUICK QUESTION: What are asteroids?

-> small rocky objects (smaller than planets) that orbit the sun.

Investigated by the probes Dawn, NEAR, Hayabusa, and OSIRIS-Rex!

Asteroids are leftover from the creation of the solar system and are typically found in between Mars and Jupiter in what is called the asteroid belt. They can be bigger than a mountain or as small as a marble, and are made of different kinds of rock. Some of the more famous ones are now either considered (or under consideration to be) dwarf planets. Examples are Ceres, Vesta, Pallas, and Hygiea.

Ida and its moon Dactyl.

As you can see in the image to the right, Ceres has many craters like our moon. Want to investigate how the speed or size of a meteor affects crater size? Go to page 34...

Chariklo has rings.

Data Dive

Iron Meteorite Minerals	Stony Meteorite Minerals
Iron 91%	Iron 26%
Nickel 8.5%	Silicon 18%
Cobalt 0.6%	Magnesium 14%
	Oxygen 6%
	Aluminum 1.5%
	Nickel 1.4%
	Calcium 1.3%

What are some differences between iron meteorites and stony meteorites?

A meteoroid is a smaller asteroid that still orbits the sun.

Whenever a meteoroid or asteroid hit the earth's atmosphere like a shooting star, they are called meteors.

If, by lucky chance a piece makes it through, the rock that has landed on earth is called a meteorite.

QUICK QUESTION: What is a comet?

-> a combination of frozen gas, rock, and dust that heat up as they get closer to the sun in their orbit. This causes a long stretch of heated gas and dust to stretch out behind them.

Comets orbit the sun in long elliptical paths. As the sun's gravity pulls them near, the energy the sun releases heats up the frozen comet releasing gas and dust.

Investigated by ICE, Deep Impact, and Stardust missions!

Make your own comet! Go to page 35...

The sun releases a solar wind which is a bunch of particles moving fast like our wind on earth. These particles help push the stream of gas and dust released by the comet always in a direction away from the sun.

DID YOU KNOW: A comet's tail can stretch more than 100 million kilometers?! That's over 60 million miles!!

A comet has parts much like our nervous system cells, although they can have different names.

Coma = outer fuzzy layer like the dendrites of the cell

Nucleus = solid inner core like the boss of our cells with the same name

Tail = stretched out gas and dust behind the comet like the long axon of the cell

Scientist Spotlight

In 1951 Dutch astronomer Gerard Kuiper proposed the existence of a ring of icy objects outside Neptune. We now know this is true and it is one of the origin areas for comets.

Would you like to go back to the solar system home page?   Go to page 2...
Would you like to start over?   Go back to page 1...

Stars and Galaxies

A star is a large spherical ball of gas that produces electromagnetic waves or radiant energy. Stars go through a whole life cycle on their own, just like us. Our sun is considered a main sequence star which is the typical star that lives a long life burning through its fuel. We have also found many unique stars that have unusual life cycles and in death can form many different objects in the universe. Travel through the universe in this section exploring the depths of space...

| Do you want to know more about the types of stars? Go to page 12... | Do you want to explore the birth and death of stars? Go to page 13... | Do you want to learn about galaxies? Go to page 15... | Would you like to discover the importance of constellations? Go to page 16... |

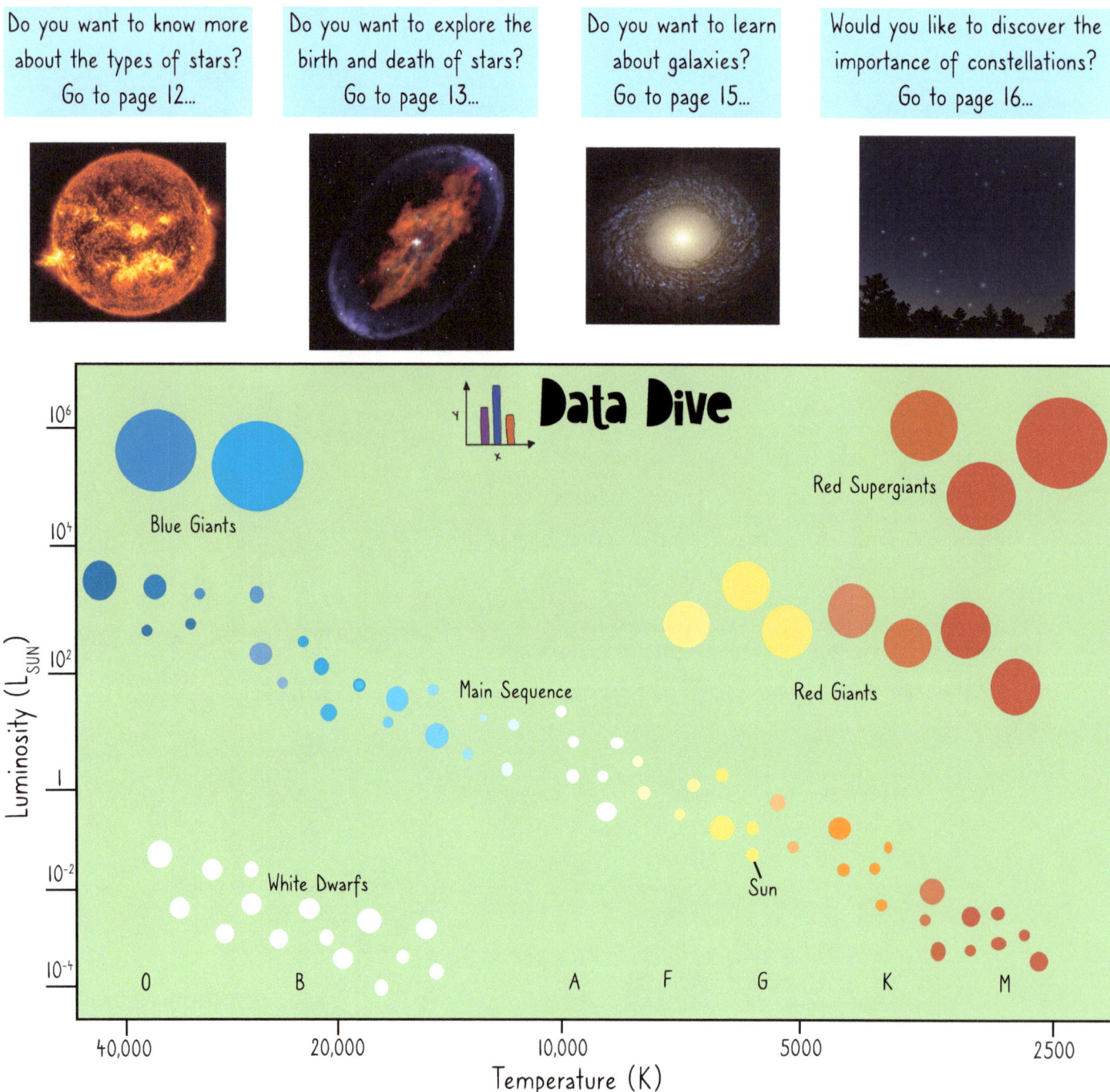

The Hertzsprung-Russell (HR) Diagram is a tool astronomers use to study the evolution of stars. It plots their brightness or luminosity against their temperature or spectral type (capital letters on the diagram). Which stars are the brightest and hottest? Which stars are bright but very cool? What spectral class is the sun?

Types of Stars

As the Physics book pointed out, there are many different wavelengths of energy coming from objects. We have invented technology that can receive the different sized waves of the electromagnetic spectrum, but they have no color. We must assign colors to the data to help us visualize what it is trying to show us. Here is an image of what the sun looks like as radio waves, microwaves, infrared waves, visible light, ultraviolet waves, x-rays, and gamma radiation.

To see the technology, Go to page 18...

To review the EM Spectrum, Go to page 20...

False Color Maps!

Stars are classified by their temperature, color, and size. Use the HR Diagram on the previous page to see these star properties.

Make and test your own star brightness detector! Go to page 35...

Blue Stars, Giants, and Supergiants

Red Giants and Supergiants

Main Sequence Yellow Dwarf

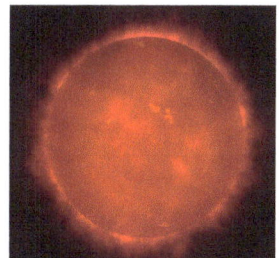
Main Sequence Red Dwarf

Stars differ in the elements that make them up and can form star systems. These star systems are a collection of two or more stars that are close together even orbiting each other in some cases.

Main Sequence Orange Dwarfs Binary (2) Star System

Star Cluster

QUICK QUESTION: What is fusion?

-> the process by which stars like our sun get power. In the reaction two atoms of hydrogen fuse together to make helium and release energy.

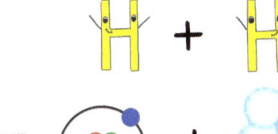
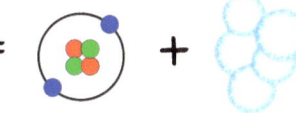

Would you like to start over?   Go back to page 1...

The Birth and Death of Stars

QUICK QUESTION:
What is a nebula?

-> a large cloud of dust and gas.

Inside nebulas, pockets of gravity form, pulling in dust and gas. As soon as enough mass or matter has been gathered in a small space (very dense), it becomes so hot that nuclear fusion starts (see previous page) and a star is born. When the fuel runs out, a couple of different things can happen depending on the mass of the star.

| Large mass stars explode as a Supernova | Small mass stars form Planetary Nebulas |

Type I Supernova
A white dwarf near another star steals hot gas from that star until an explosion occurs.

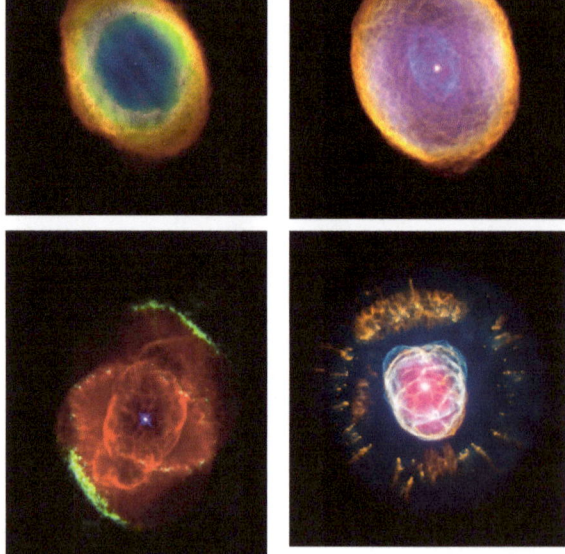

Type II Supernova
A large star runs out of fuel and collapses inward blasting out energy and matter much faster than the speed of light.

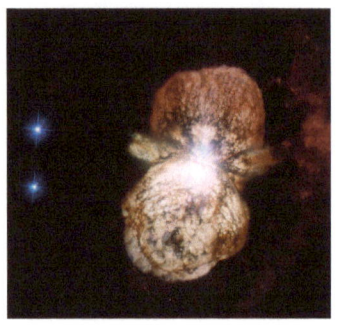

Supernova Remnants are the remains of the explosion from a supernova!

 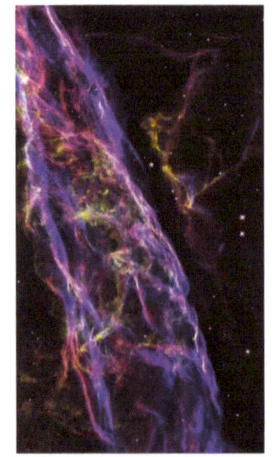

CREATIVE CORNER

Create your very own Supernova FlipBook! Share an animation or video to be featured on the HTH Space book website at submissions@howsthathuman.com

13

Planetary nebulas eventually cool down into white dwarfs. White dwarfs are about as big as the earth but with the mass of the sun crammed inside. That's SUPER dense! White dwarfs eventually fade into brown dwarfs and then black dwarfs.

After a supernova explosion, the gases and dust become a new nursery for birthing stars. Neutron stars form and can evolve into pulsars, or a black hole forms.

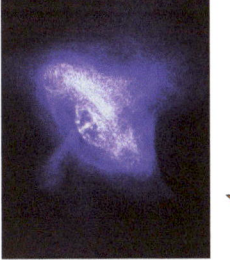

QUICK QUESTION: What is a neutron star?

-> the remains of a high mass star.

QUICK QUESTION: What is a pulsar?

-> a spinning neutron star.

Scientist Spotlight

Northern Ireland's astrophysicist Jocelyn Bell Burnell discovered the first radio pulsars in 1967. She recognized that an object was giving off regularly repeated radio waves.

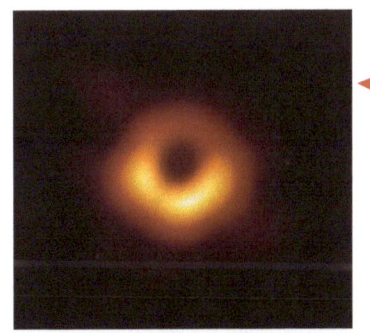

This first image of a black hole was taken by the Event Horizon Telescope collaboration from around the world. Black holes have such extreme gravity that not even light can escape their pull. Scientists use X-rays to study the energy around black holes, and recently captured rings around a black hole.

Ready to Research?!

What would happen if you fell into a black hole?

Search Suggestion: falling into a black hole

Would you like to start over? Go back to page 1...

Would you like to go back to the stars and galaxies home page? Go to page 11...

Would you like to see the telescopes that have taken all these amazing photos? Go to page 19...

Galaxies

QUICK QUESTION: What is a galaxy?

-> a huge group of stars, gas, and dust that are held together by gravity.

Make your own galaxy! Go to page 36...

Spiral Galaxy
Has a bulge in the middle and arms that spiral out like a pinwheel.

Elliptical Galaxy
Looks like a large ball or flattened ball, much like a football.

Irregular Galaxy
Does not have a regular shape, odd-looking.

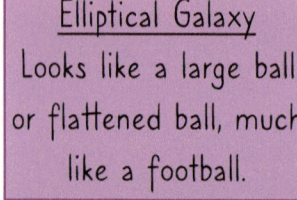

Would you like to go back to the stars and galaxies home page? Go to page 11...

DID YOU KNOW: There are billions of galaxies in the universe and some have over a trillion stars each?!

Would you like to see the telescopes that have taken all these amazing photos? Go to page 19...

Constellations

Constellations have played a part in human history for thousands of years. Their predictable movement allowed for calendars to be created and agriculture to be successful. Their shapes gave rise to myths and creation stories. Ultimately, using them for navigation led to the exploration of the entire world.

Track stars with an astrolabe like your ancestors! Go to page 36...

CREATIVE CORNER

What's Your Sign? Research the astrological sign for your birth date and find the constellation that matches. Get creative and make a 3D version of the constellation to share on the HTH Space book website. Submit to submissions@howsthathuman.com

Currently, we use them for naming purposes so that scientists from around the globe can communicate, naming the brightest star in the constellation first down to the dimmest star last, using the Greek alphabet.

Would you like to start over? Go back to page 1...

Space Exploration

Humans have always looked to the skies; dreaming, singing songs, writing stories, etc. What is out there? Is there another intelligent life? What can we learn from space? Can we travel in space? How can we address the problems we face in space?

How do stars explode?

How far have we discovered in space?

How do planets keep their circle?

How long do we survive in space?

Have aliens ever visited earth?

Do you want to learn about the types of technology used to explore space? Go to page 18...

Scientist Spotlight

Thebe Medupe is an astrophysicist from South Africa who wants to encourage interest in his field. He started Astronomy Africa and promotes the appreciation of cultural beliefs to understand the universe.

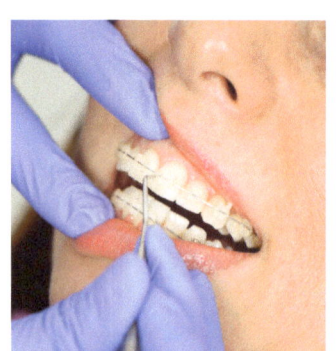

Do you want to see some examples of human advancements and space exploration spin-offs here on earth? Go to page 25...

Do you want to examine the issues humans face in space? Go to page 27...

Technology

Scientist Spotlight

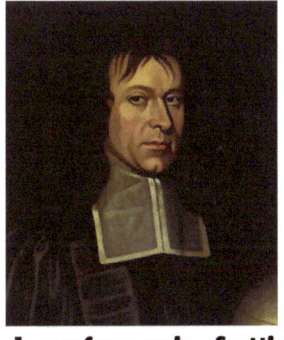 

James Gregory is a Scottish astronomer who designed the Gregorian reflective telescope. His measurement of the distance from the earth to the sun became the basis of the Astronomical Unit (AU), a measurement we use today for distances in space.

Telescopes
Go to page 19...

Rovers
Go to page 22...

Satellites
Go to page 21...

Manned Missions
Go to page 23...

Deep Space Network
Go to page 24...

18

Telescopes

The atmosphere plays a critical part in protecting life on earth by absorbing or reflecting the harmful waves of radiation on the electromagnetic spectrum. This has affected the way we study astronomy and led to the invention of some pretty incredible telescopes around the world.

QUICK QUESTION: What is radiation?

-> energy that comes from electromagnetic waves. The waves begin at a source and travel through space at the speed of light.

Check out the HTH Physics book for more on waves and the EM spectrum.

Ground-based telescopes can analyze radio waves, some infrared waves, and visible light. We had to create space telescopes in order to analyze microwaves, more infrared waves, ultraviolet waves, x-rays, and gamma rays because of our atmosphere. In space, we get a much clearer view of the universe anyhow! Here are some famous ground radio telescopes around the world....

FAST
The world's largest single-dish radio telescope located in China.

LBT
A unique, very large telescope with a pair of mirrors that is located in Arizona.

SALT
A large optical telescope located in South Africa.

Keck 1 and 2
A pair of sophisticated optical and infrared telescopes on Mauna Kea in Hawaii.

VLT
Four individual optical and infrared telescopes located in Chile.

Gran Telescopio Canarias
The world's largest single-aperture optical telescope located in Spain.

Scientist Spotlight

Australian Ruby Payne-Scott was the first female radio astronomer. Her work in radio astronomy laid the foundation for the discovery of black holes, pulsars, and understanding solar storms and flares.

There are two types of telescopes.

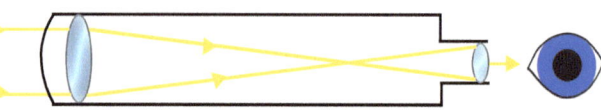

Refracting
A lens focuses the light to an eyepiece which then magnifies the image.

Reflecting
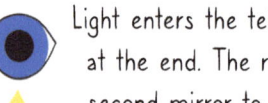
Light enters the telescope and reflects off a mirror at the end. The reflection is then bounced off a second mirror to the eyepiece for magnification.

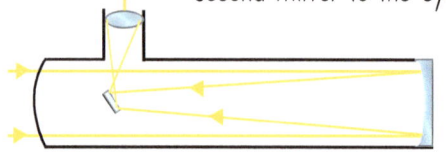

Check out photos by these telescopes! Go to page 11...

Image Examples: Crab Nebula | Wavelength Type and Size | Telescope Examples

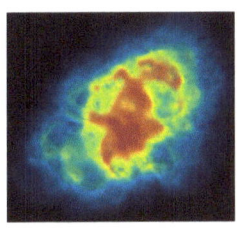

RADIO

VLA
The Very Large Array consists of 28 radio telescopes in New Mexico.

Cosmic background radiation or cosmic microwaves are the oldest EM waves on the spectrum. They are studied to understand the early universe.

COBE
The Cosmic Background Explorer was a satellite dedicated to microwaves and understanding the Big Bang Theory.

INFRARED

Spitzer
A space telescope analyzing the infrared waves in our galaxy and beyond. Was the first to detect light from a planet outside the solar system.

VISIBLE

Kepler
A space telescope with the mission of finding exoplanets, especially ones in habitable zones.

ULTRAVIOLET

Hubble
This multi-talented space telescope has given us stunning images and information from infrared, visible and ultraviolet waves!

X-RAY

Chandra
This space telescope detects x-rays from very hot regions in space giving us data on black holes and exploding stars.

GAMMA RAY

Fermi
A space telescope that investigates gamma rays, the fastest moving waves with the highest energy.

Would you like to go back to the Technology home page?   Go to page 18...
Would you like to go back to the space exploration home page?   Go to page 17...
Would you like to start over?   Go back to page 1...

Satellites

QUICK QUESTION: What is a satellite?

-> a man-made object, moon, or planet that orbits another planet or star. Natural satellites are like the moon orbiting the earth or the earth orbiting the sun.

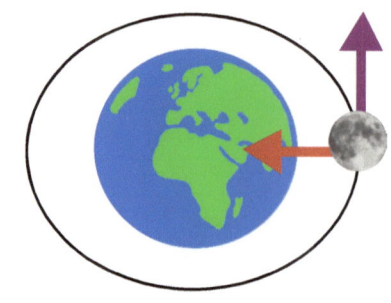

There is a lot of physics behind achieving orbit! The satellite with its specific mass has to be traveling fast enough to balance the pull of gravity from the larger object. It becomes a constant tug-of-war. Make sure to check out the HTH Physics book for more information on forces and motion.

Investigate how differences in satellite speed and the gravitational pull of a planet affect the orbit of a satellite, Go to page 37...

Geostationary satellites revolve around the earth in 24 hours, making it look like they stay in the same position all day.

Polar satellites orbit the earth in a north-south direction, capturing data of the entire world in one day.

SES, communications

LANDSAT, earth science

GOES, weather

NOAA-POES, weather and climate, specifically short term and long term weather forecasting

NAVSTAR and GPS, navigation

Special Note: International Space Station (ISS), microgravity and space research facility with 5 countries collaborating

Special Note: TERRA, climate and environment, orbits with the sun

Would you like to go back to the Technology home page? Go to page 18...

Ready to Research?!

What is escape velocity?

Search Suggestion: escape velocity, earth escape velocity

DID YOU KNOW: Sputnik, the first satellite launched into space, was the size of a basketball and weighed 183 pounds?!

Rovers

QUICK QUESTION: What is a rover?
-> a man-made planetary surface exploration robot.

Make your own rover!
Go to page 38...

What do we face as humans in space?
Go to page 27...

The 5 rovers of NASA that have successfully landed on Mars are shown below. These rovers drive around taking pictures of and collecting samples on the surface of the planet. We can use the information they send back to understand more about the planet and what we face exploring and colonizing it in person.

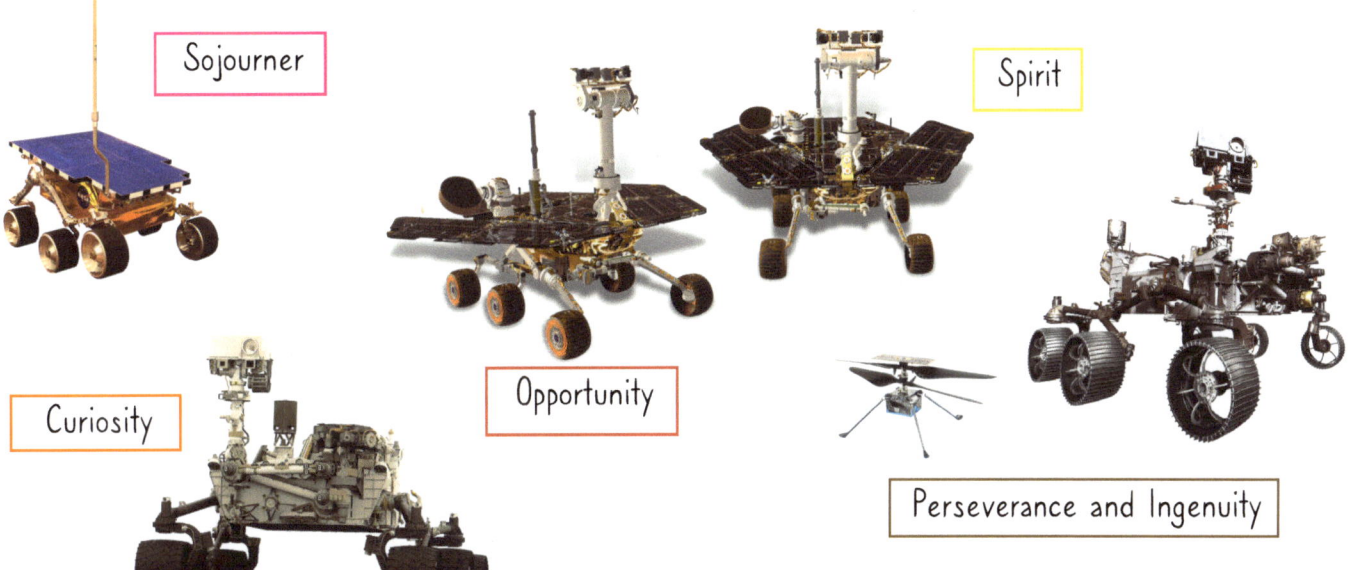

Sojourner

Spirit

Opportunity

Curiosity

Perseverance and Ingenuity

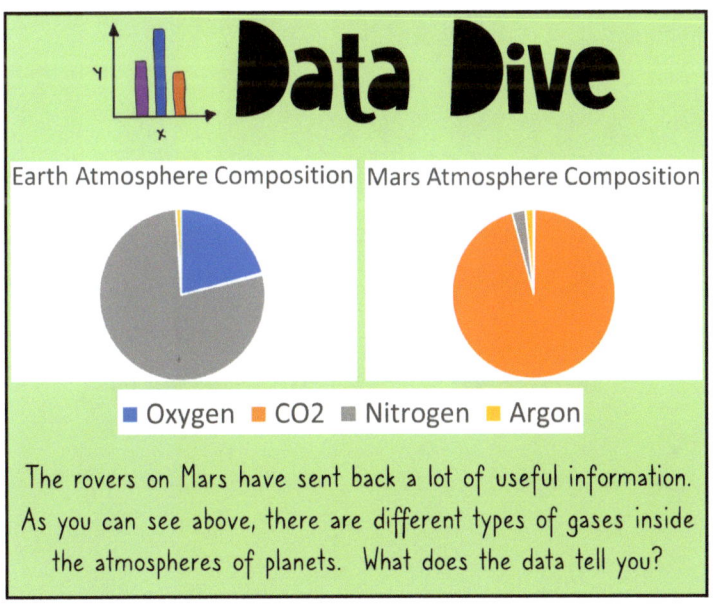

Data Dive

Earth Atmosphere Composition | Mars Atmosphere Composition

■ Oxygen ■ CO2 ■ Nitrogen ■ Argon

The rovers on Mars have sent back a lot of useful information. As you can see above, there are different types of gases inside the atmospheres of planets. What does the data tell you?

Rovers are a lot like you!

Body	protects the "organs"
Brains	processes information
Neck and Head	supports the camera
Arm	increases reach
Legs	wheels for movement
Eyes and Senses	camera and data gathering instruments
Communication	antennas for "speaking"
Temperature Control	heaters and insulation
Energy	solar panels and batteries

Would you like to go back to the space exploration home page?   Go to page 17...
Would you like to start over?   Go back to page 1...

Manned Missions

Since the first man made it to space, cosmonaut Yuri Garagin, many different programs have been developed worldwide to increase our knowledge of the universe. Below are the programs the U.S. put into play as humans achieved spaceflight.

Mercury Program
Put the first American astronaut, Alan Shepard, into space.

Gemini Program
Practiced space rendezvous and spacewalks.

Scientist Spotlight

Mary Golda Ross was the first known Native American female engineer. She was the first female engineer to work for Lockheed. Her research led to the advancements in space flight put to use in the Apollo Program.

Apollo Program
Put the first man on the moon and saw six total moon landings over the years. It highlighted human perseverance with the Apollo 13 mission.

Skylab
The first American space station.

Artemis Program
The current program to get humans back to the moon again.

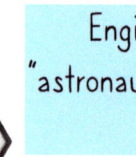

Engineer a safe vehicle for an "astronaut" or blast off your own rocket, Go to page 38...

Space Shuttle Program: this program reused spacecraft to shuttle cargo and astronauts to space and back for maintenance of other equipment or research aboard the ISS.

Would you like to go back to the Technology home page?   Go to page 18...
Would you like to go back to the space exploration home page?   Go to page 17...

Deep Space Network

The Deep Space Network, or DSN, is NASA's way of communicating through space using large radio antennae. There are three locations around the globe where groups of radio antennae are positioned so that information from spacecraft can be received at all times of the day and night. This allows for communication, navigation, and scientific research.

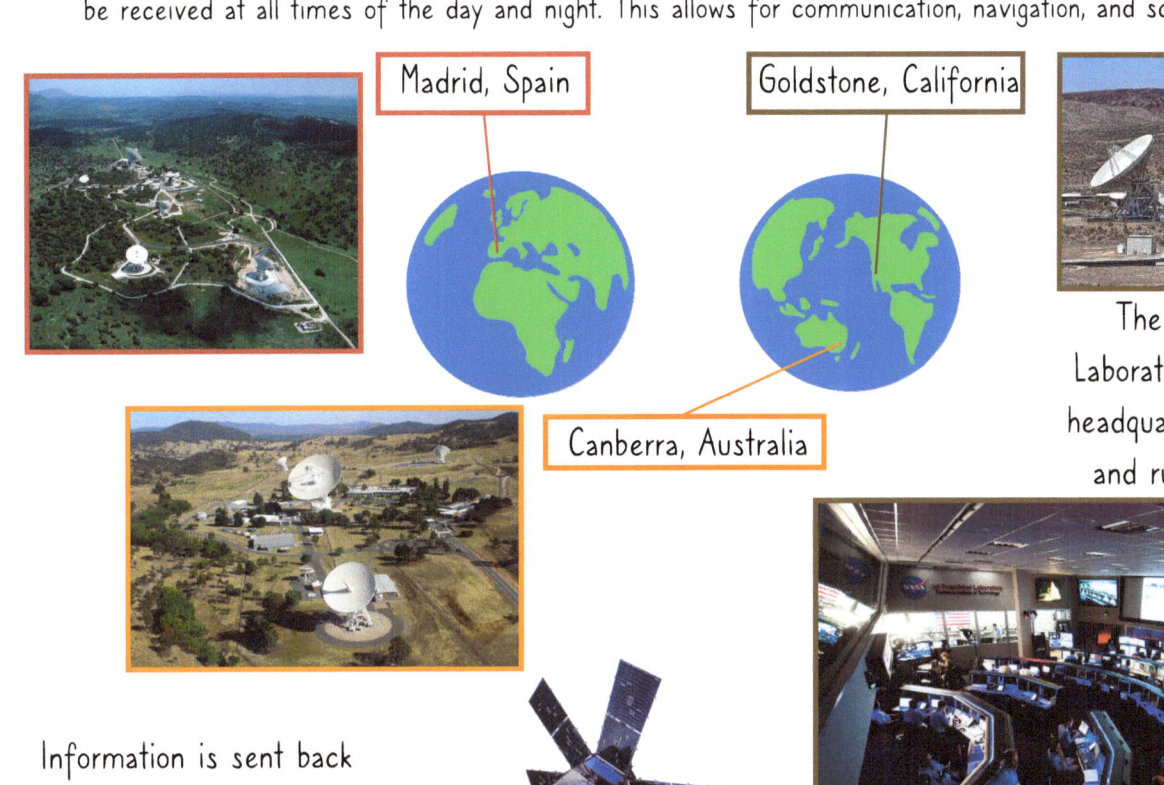

Madrid, Spain

Goldstone, California

Canberra, Australia

The Jet Propulsion Laboratory, or JPL, is the headquarters for the DSN and runs this location.

Information is sent back and forth between the satellite and the antenna.

Test your own ears with a radio "antenna", Go to page 36...

If you could text an alien, what would you say? Share your message to submissions@howsthathuman.com to be featured on the HTH Space book website!

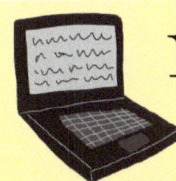

Ready to Research?!

How will we communicate with Mars?

Search Suggestion: communications with earth, mars communication, Deep Space Network and Mars

DID YOU KNOW: Voyager 1 satellite has left the solar system and it takes 20 hours for the signal to reach earth?!

Would you like to start over?   Go back to page 1...

Human Advancements

SPACE ROBOTICS HAVE INSPIRED THE DESIGNS FOR EXOSKELETONS ALLOWING PARAPLEGICS TO WALK!

Telemedicine has become more and more needed and with the advent of the Advanced Diagnostic Ultrasound in Microgravity (ADUM), people in remote areas will be given access to highly technical medical information easier.

On Mars, there are no fossil fuels; so we have to rely on renewable resources. A new wind turbine has been designed to survive in harsh conditions, such as low temperatures. Currently, there are over 800 of these wind turbines generating electricity in the south pole.

Wireless headphones were created during the Apollo program and, as you probably already know, are widely used in business and pleasure.

Remember Neil Armstrong's famous words the next time you play video games! This technology has been around for a while!

Would you like to go back to the space exploration home page?   Go to page 17...

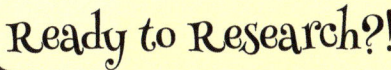

Ready to Research?! Look up more examples!
Search Suggestion: benefits of space exploration, space spin-offs

Light-emitting diode (LED) chips were invented to help grow plants in space. They are more energy-efficient and do not release as much heat as normal light bulbs. Turns out they have an additional medical benefit when combined with drugs that attack tumors destroying cancer cells.

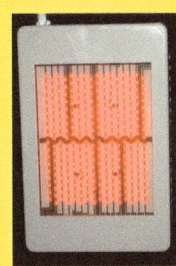

Monitoring the health of astronauts is very important and through the development of different devices, we now have implantable heart monitors.

To think about how microgravity would affect the human heart, we now have everyday technology here with gravity.

The athletic shoe was a result of research into shock-absorbing materials.

To think you have been walking in the shoes of the astronauts!

The need for filtering and recycling water on manned missions drove the development of water purification systems that are used worldwide saving lives.

Would you like to start over?   Go back to page 1...

Humans in Space

The Human Research Program at NASA has been studying what happens to the human body in space for decades. Their valuable research has led to the invention of new strategies, devices, and procedures to keep astronauts not only safe but healthy both mentally and physically.

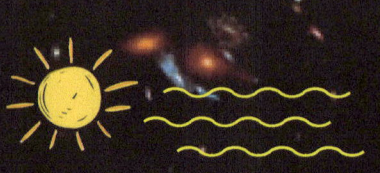

Radiation
Go to page 28...

Distance From Earth
Go to page 30...

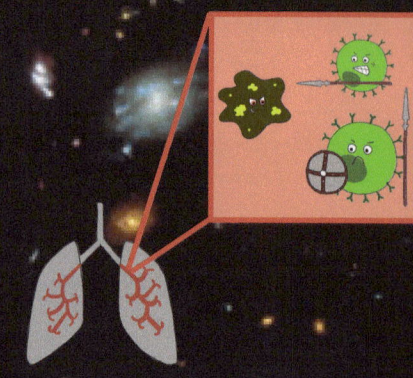

Isolation and Confinement
Go to page 29...

Gravity
Go to page 31...

Closed Environment
Go to page 32...

Which of these would you struggle with the most? Take the poll on the HTH Space Book website!

CREATIVE CORNER

The food astronauts eat is specialized for many reasons. Think about a recipe that you would like to see in space and share it at submissions@howsthathuman.com to be featured on the HTH Space book website!

SPECIAL Spotlight!

Lien Pham immigrated from Vietnam and became a seamstress in America. Little did she know that she would end up designing thermal blankets for missions to Mars!

Radiation

The Causes
- Particles trapped in the Earth's magnetic field
- Solar particles from the sun
- Galactic cosmic rays
- The amount of radiation in space is greater than on earth
- The type of radiation exposure will be different from on earth

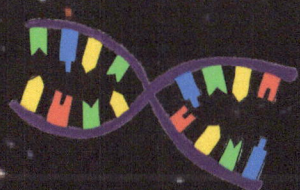

Thanks magnetic field!

Test the effects of solar radiation at home by making your own solar oven! Go to page 36...

The Effects
- Degenerative Diseases (heart disease, cataracts, digestive and respiratory disease)
- Cancer
- Radiation sickness
- Changes in the nervous system

Radiation damages your DNA causing your cells not to work right anymore.

Real-World Solution Example
The Orion spacecraft will be what brings humans back to the moon as part of the Artemis Program. Orion will be equipped with a radiation-sensing instrument that will alert astronauts to take shelter. The crew will then take shelter in the center of the crew module where storage bags exist. Adding all the extra mass around them forces the radiation to pass through more layers adding protection. The storage bags will conveniently contain food and water.

The Solutions
- Monitoring of health
- Medicines
- Healthy diet
- Radiation shielding
- Specific Operating Procedures
- Performing research in ground-based facilities

Would you like to go back to the space exploration home page?   Go to page 17...
Would you like to start over?   Go back to page 1...

28

Isolation and Confinement

The Causes
- Limits to the weight of spacecraft
- Limits to the speed of space travel
- Limits to crew number
- Crew members from different backgrounds
- Small and noisy environment
- Heavy workloads
- Shifting schedules
- Boredom
- Missing loved ones
- Guilt
- Less contact with outside people
- Small spaces

Something To Think About
The mission to Mars will put humans in space for longer than we ever have been before. They will be isolated from family and friends, confined to small private spaces, and will be a multi-cultural crew. We must train the right crew; with team dynamics and cultural sensitivity a priority for the success of the mission. NASA research has pointed out that both the length of the mission and the type of isolated and confined experience matter.

The Effects
- Sleep problems
- Fatigue
- Mood decline
- Behavioral changes
- Sensitivity
- Stress
- Psychiatric disorders

Get quality sleep!!

HI-SEAS
The Hawai'i Space Exploration Analog and Simulation research station is testing long missions in space by allowing astronauts to "live" many months as they will in space. The crews perform the usual "spacewalks" in suits on the side of Mauna Loa and perform scientific research. We are learning a lot from these simulated missions on how humans deal with isolation and confinement, and what we can do to make it better.

The Solutions
- Self-assessments
- Light technology
- Gardening
- Journaling
- Virtual Reality sessions
- Learn something new (language, skill)
- Develop strategies to get the best combination of crew members

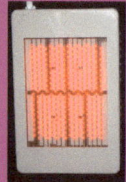

Would you like to go back to the humans in space home page?   Go to page 27...

Distance From Earth

The Causes
- No pharmacy access
- No grocery stores
- Limited storage
- Limited weight
- Communication delay
- No regular receiving of supplies

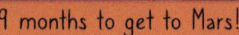 9 months to get to Mars!

3 days to get to the moon!

20 minutes to message Mars!

Dehydrated foods and drinks (water has been removed) are mostly used in space because they don't weigh very much. Also, as a result of using fuel cells in space, water is created and can be used to rehydrate the food.

The Effects
- Food storage challenges
- Lack of medical care
- Ineffective medications
- Equipment failure

Want to see more space spin-offs? Go to page 25...

Extra Benefits!
In researching methods for better nutrition while keeping weight down on the spacecraft and dealing with limited space, researchers discovered a way to naturally produce an important biological compound found in breast milk. This compound is now used in over 90% of infant formulas here on earth.

The Solutions
- Food nutrient preservation
- Medicine packaging and preservation
- Sustainable food systems
- Virtual assistants
- Clinical decision support tools
- Medical Training (IV, ultrasound, lab tests, procedures)

Would you like to go back to the space exploration home page? Go to page 17...
Would you like to start over? Go back to page 1...

Gravity

The Causes
- Up to 3 different gravity fields (Earth, space, Mars)

Mars gravity is 1/3 of Earth's gravity

Real-World Research Example
The astronaut Twin Study that compared Scott Kelly's body after a year in space to his twin brother Mark Kelly's body here on Earth, illustrated the drastic changes space has on the body.

The Effects
- Reduced muscle mass
- Bone loss
- Fluid shifts
- Eyesight issues
- Changes in senses
- Coordination and balance
- Motion sickness
- Lightheaded and fainting

Experiment with the human body! Go to page 39...

1. The ends of his chromosomes gained length in space but quickly returned to normal back on earth.
2. He was the first person vaccinated while in space, and his immune system responded appropriately.
3. Certain genes were turned on or off, and although some returned to normal others did not after he was back on Earth.
4. His mental abilities remained unchanged, however, he had less speed and accuracy after returning to Earth for 6 months.
5. He lost body mass and gained minerals due to the controlled diet and regulated exercise.
6. The beneficial bacteria in his gut drastically changed while in space, but returned to normal when he was back.
7. His carotid artery thickened while in space and after landing giving us data on the circulatory system.
8. He had extra fluid in his eyes and further data will help researchers understand the vision problems in spaceflight.

All this research into astronauts will teach us about disease risk factors here on Earth!

The Solutions
- Exercise
- Medications
- Pressure devices (compression cuffs)
- Fine motor testing
- Ultrasounds
- Fitness evaluations
- Virtual workout partners
- New treatments
- Preventative Measures

Would you like to go back to the humans in space home page? Go to page 27...

Closed Environment

The Causes
- Microorganisms that live on the human body transfer more easily between people in a closed habitat
- Stress hormones are elevated and alter the immune system., increasing susceptibility to illness or allergies
- Not comfortable in living quarters

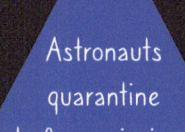

Astronauts quarantine before missions

The Effects
- Temperature changes
- Exposure to contaminants
- Celestial dust exposure
- Altered immune system
- Autoimmune diseases
- Stress

The immune system has many cells ready to fight, but that doesn't mean they will win. Perhaps a new organism from a different planet will enter our bodies undetected. Or, because of poor health, the immune system doesn't have enough energy to fight. It is extremely important that we put astronauts in space with the best chance of avoiding disease from the start.

Want to see more space spin-offs? Go to page 25...

Temperature Balance
When designing spacecraft we must consider the extreme temperatures in space. The sun's radiation will reach temperatures of 250 degrees F while the opposite side freezes at negative 250 degrees F. Thermal (heat) balance is always a priority and one way we solve this issue is by using a highly reflective insulation called Multi-layer Insulation. One layer called Mylar is used in thermal blankets here on earth for camping or shock victims.

The Solutions
- Thermal control systems
- Air quality monitoring
- Water treatment
- Routine cleaning
- Air filter maintenance
- Immunizations and quarantines before flights
- Earth-like LED lights
- Regular microbe analysis swabbing of body and spacecraft
- Carefully planned living quarters and work environments

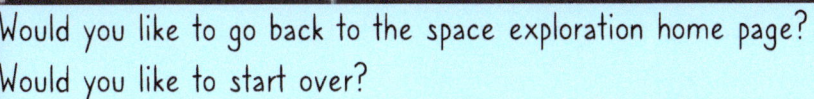
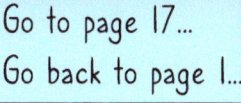

Would you like to go back to the space exploration home page? Go to page 17...
Would you like to start over? Go back to page 1...

Hands-On From Home

The following experiments and activities can be used to model, simulate, or explore astronomy. Have fun!!! Example experiment videos are on the HTH Space Book website.

Feel free to share any of your experiments at submissions@howsthathuman.com to be featured on the Space Book Science Fair page.

Scaling The Solar System

In this activity, you will make a model of the size of the planets in our solar system as well as the distance between the planets. Scientists use models all the time in place of the real thing, for example, the sheer size of the solar system is something we cannot create in a lab!

For more on the solar system, Go to page 2...

Step 1 – Make Your Planets!
1. Here is where you can get a little creative. You can make paper versions of the planets and color them (some like Jupiter are quite large so you may have to tape paper together). Or you could find circle-shaped objects of various sizes around your house like plates or sports balls to measure and hope they match in size. You could also create circles from string and make circle shapes on the floor or table, or draw circles with sidewalk chalk outside.
2. A scale is used as a conversion of one thing into another. You may be taking something small and making it larger, or in our case, the reverse. We are taking something very large and making it much smaller. In this step, you are using a scale to determine the size of the planets that you can easily recreate. Using the given scale factor Earth diameter = 4 cm, fill in the sizes of the other planets in the chart and then make them or measure them using the items from your home and a ruler.

A. Set up two fractions equal to each other.
B. Cross multiply and set those answers equal to each other.
C. Divide the whole number by the number multiplied by "?" and that is your answer!

PLANET	DIAMETER (km)	SCALE DIAMETER (cm)
Sun (star)	1,500,000	420
Mercury	5000	
Venus	12,000	
Earth	13,000	4
Mars	7,000	
Jupiter	143,000	
Saturn	120,000	
Uranus	51,000	
Neptune	50,000	
Pluto	2,000	

A
$$\frac{420}{1,500,000} \times \frac{?}{13,000}$$

B
$$420 \times 13,000 = ? \times 1,500,000$$

C
$$5,460,000 / 1,500,000 = 3.6 \text{ or } 4\text{cm}$$

Step 2 – Set Up the Solar System!
1. The AU or astronomical unit is a set measurement based on the distance from Earth to the Sun. 1 AU is 93 million miles!! We are using the scale factor of 1 AU = 10 cm and given the measurement is starting from the sun, the sun will have a value of 0 cm. Use the bottom of the page to complete the calculations for the table.
2. Once you have completed your calculations, it is time for you to line up your final solar system! Using your planets from Step 1 and a ruler/meterstick, start lining up your planets in order using the scale value distance from the sun in the table (pick something to act as the sun or starting point).

OBJECT	AU	SCALE VALUE (cm)
Sun (star)	0	0
Mercury	0.4	4
Venus	0.7	
Earth	1.0	10
Mars	1.5	
Asteroid Belt	2.8	
Jupiter	5.2	
Saturn	9.6	
Uranus	19.2	
Neptune	30.0	
Pluto	39.5	

Earth-Moon-Sun Relationship For more info, Go to page 3...

We can make our very own model of the moon revolving around the earth WHILE the earth revolves around the sun!

Materials: 1-2 sheets of paper, scissors, colors (crayons, markers), and 2 paper fasteners (brad, brass fastener)

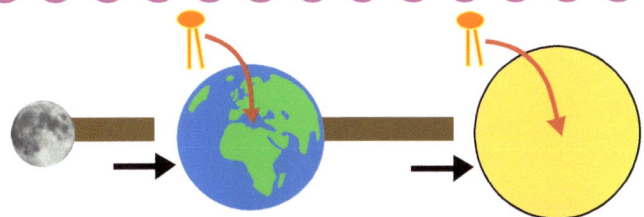

Procedures:
1. Create 3 different circles on your paper to represent the earth, moon, and sun and color them. Try and keep size in mind even though this won't be to scale. Leave room for a 2-inch long thin rectangle strip to extend from the moon, and a 4-inch long similar strip extending from your earth.
2. Poke a paper fastener through the middle of the earth and through the end of the 2-inch strip hanging off the moon and bend it to keep the two together. Now you can revolve your moon around the earth.
3. Poke the second paper fastener through the middle of the sun and then through the end of the strip hanging off the earth and bend it again. Now you can revolve the earth around the sun.

Moon Model

Time to model the phases of the moon and the two types of eclipses!

Materials: large ball, small ball, light source

Procedures:
1. Collect two different-sized balls, one to represent the earth and the other to represent the moon. Grab yourself a light source like a lamp or flashlight to act as the sun.
2. In a dark or semi-dark room, place the earth ball in front of the moon and shine the light at the earth. What happens to the amount of light that reaches the moon? What type of eclipse is this?

3. Keeping the light and the earth in the same position, move the moon in between them. What do you observe on the earth? What type of eclipse is this?

For moon phases, Go to page 4...

4. Model the phases of the moon by leaving your light source and earth in the same position, but move your moon ball around the earth. Look for the amount of light reflecting off the surface of your moon as you move from what would be your position on earth.

Craters!

Does speed or size matter more when a meteorite hits a planetary body?

Materials: several sticks of clay, a large pan/tub/box, baking flour, ruler/yardstick

Procedures:
1. Pour your baking flour into your large tub and smooth the top as best as you can.
2. Make 3 different balls out of clay; the first should be half a stick, the second 1 stick, and the third 1.5 sticks (use the other half leftover from your first ball).
3. To answer the question, first, drop all three balls from the same height, one at a time recording the size of the crater (measure across the middle from edge to edge, this is called the diameter) and smoothing the surface before each drop. This looks at how the size will affect crater size.
4. Next, pick one of the balls to use and drop it from 3 different heights, such as 10cm, 20cm, and 30cm. Don't forget to smooth the surface before each drop. As you change the height you are changing the speed, increasing with height. What does your data say about the speed of a meteorite? Can you tell if speed or size matters more? For more on meteorites, Go to page 9...

Make Your Own Comet!

For more on comets, Go to page 10...

Materials:
- Protective fabric gloves and eyewear
- 1 cup water
- 1 cup potting Soil
- 1 tablespoon soy sauce
- 1 one-gallon plastic storage Bag
- Light bulb/Hot light source

PARENTAL SUPERVISION BELOW
- 0.25 - 0.5 lbs dry ice, broken into small pieces (Do NOT handle without protective gloves!)
- 1 tablespoon ammonia (Exposure may result in irritation of eye, nose, and throat.)

Procedures:
1. Put on the fabric gloves and protective eyewear.
2. In the plastic storage bag:
   a. pour 1 cup of water
   b. pour 1 cup of potting soil
   c. carefully add 1 tablespoon of ammonia (Parental supervision required!)
   d. add 1 tablespoon of soy sauce
   e. carefully add a handful of the broken-up dry ice to the bag (Parental supervision required!)
3. Smash and mix the materials in the bag into a comet clump.
4. Once your comet has formed, you can remove it from the bag (still wearing your protective gloves).
5. Hold your comet in front of your light source, aka the sun. Watch what happens to its tail, why does this occur?

Star Brightness Detector!

You can construct your own tool to measure and categorize stars based on their brightness which depends on how much light the star is putting off and how far it is from Earth.

For more on stars, Go to page 11...

Materials: cardboard, ruler, scissors, colored cellophane paper, tape, and a clear night sky

Procedures:
1. In a piece of cardboard, cut four 1.5 inch wide rectangles in a column with space in between each.
2. Over all four rectangles, tape one sheet of cellophane.
3. Over the last three rectangles, tape another sheet of cellophane (you can cut to fit and save the piece for the last rectangle.
4. Over the last two rectangles, tape another sheet of cellophane.
5. Tape one final overlapping sheet of cellophane on the last rectangle only (the piece you saved from earlier would work here.)

6. Head outside and view the night sky. Using the tool, count how many stars you see in the first rectangle with one sheet. Look through the other rectangles and count how many stars you can see with two sheets, three sheets, and four sheets. What trend do you see? In which rectangle could you see the most stars? Why do you think this is?

Your Very Own Milky Way Galaxy

Time to get creative and make your personal version of the milky way spiral galaxy! *For more on galaxies, Go to page 15...*
1. Grab cotton balls, a paper plate, glue, and whatever materials you want to decorate with (paint, sparkles, markers, crayons).
2. Tease apart some cotton balls to make half circle domes on the top and bottom of the paper plate and glue them in place.
3. Decorate the entire plate and cotton balls.
4. Use scissors to cut the spiral design out of the plate. Don't forget to label our location!

Step Back In Time

For centuries of navigation, humans have used an astrolabe. An astrolabe measures the position of a star in the sky using the horizon line as a baseline.
How does this relate to constellations? Go to page 16...
Materials: a protractor, string, scissors, a weight (washer, rock), and a night sky
Procedures:
1. Cut a 12-inch piece of string and tie a weight to one end. Tie the other end to the hole in the middle of the crossbar on the protractor.
2. Flip the protractor so that the zero-degree mark is near your face and the curved part is down with the string and weight hanging.
3. Sit on the ground with the end of the protractor near your eye and look down the edge at a star you want to measure.
4. Hold the string against the protractor and note which degree the string crosses on the curved part of the protractor. This is how far above the horizon the star is.
4. Take measurements of the star every 30 minutes at least 2 more times. What trend do you see?

Solar Oven

Put radiation to use when creating your own solar oven!
For more on radiation, Go to page 19...
1. There are several ways to make your oven, but a few things will stay the same. Whether you recycle a chip canister with reflective foil on the inside, or line a cardboard box with tin foil, you must angle and reflect the sun's rays towards your food inside your "oven".
2. You will need to have an opening for the rays to enter, and you can seal it off with plastic wrap. This traps the heated gas molecules inside much like greenhouse gases.
3. A thermometer can be placed inside the oven to monitor the temperature.
4. Have fun trying different designs, think about insulation (newspaper) and using black paper as well (absorbs heat the most). Try different foods; you can even run a skewer through the sides of your oven!

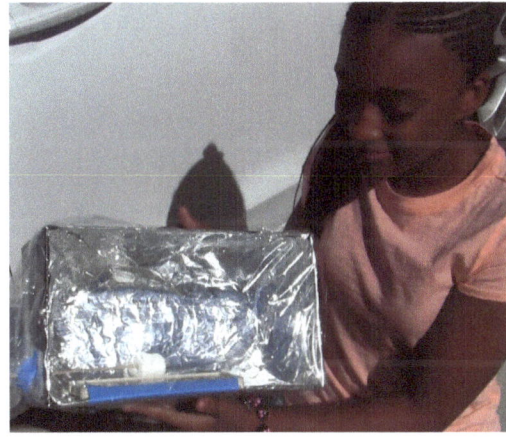

Just A Super Large Funnel

Radio antenna are shaped the way they are for a reason, test it out yourself! *How does this relate to the DSN? Go to page 24...*
1. Get a poster board or very large paper and roll it into a cone shape. Tape it in place and put the small end to your ear. Test it inside and outside, what do you hear? What does this tell you about the shape?

Satellite Investigations

In orbit, the speed of the satellite and the gravitational pull of the larger planetary body nearby matters in order to maintain a steady orbit. If the gravitational pull is not strong enough and speed too fast, the satellite will fly off into space. If the pull of gravity is too strong and the speed too slow, the satellite will crash into the planet.

Materials: disposable gloves, cookie sheet, 2 sheets of finger paint paper, 1 marble, 2 different colors of food coloring, cardboard tube (paper towel tube), tape, modeling clay, plastic lids or dishes, ruler

Procedures:

For more on satellites, Go to page 21...

1. Tape one finger paint paper to your cookie sheet.
2. Using your ruler, you will make 2 sets of stilts for the cookie sheet. One set needs to be 3cm tall, the second 8cm tall. Try and make them rectangular blocks. Place the 3cm set of stilts under one side of the cookie sheet causing it to lean at an angle. Changing this lean from 3 to 8cm will represent the increase in the gravitational pull of a "planet". Perhaps Earth versus Neptune? You decide!
3. Using more clay, you will build a support for the cardboard tube. You will want one end of the cardboard tube to rest at the end of one of the sides near the top of the leaning cookie sheet. This will represent the speed of the satellite, aka the marble. Position the tube so that the bottom of the other end not touching the cookie sheet is 10cm tall. You may use clay and tape to help secure it in place. You will also be increasing the speed by repositioning the tube such that the bottom edge will end at 20cm, so keep that in mind.
4. To observe the path of the satellite, pour about 15 drops of each color of food coloring on separate plastic lids or dishes. Make sure you have your disposable gloves on at this point.
5. At the 3cm stilted cookie sheet level (Earth) and 10cm tall tube (slow speed), roll the marble in one food coloring and release it down your tube. You may want to repeat a few times to make sure the color shows up on the paper.
6. Reposition the end of the tube to the 20cm height. Making sure the marble is clean, roll it in a different food coloring and release it several times from this new height. What happens to satellites when you increase speed?
7. Repeat steps 5 and 6 using the second sheet of finger paint paper and the cookie sheet with the 8cm stilts. What happens to satellites when you increase the gravity of the planet?

Racing Rovers

Here is a nice design challenge to test your rover skills!

For more on rovers, Go to page 22...

Materials: corrugated cardboard, tape, 2 rubber bands, 2 smooth hard candies with holes, scissors, 1 pencil, 1 straw, and a ruler

Procedures:

1. Make the rover body by bending down the sides of a piece of cardboard. Poke axle holes evenly with your pencil towards the front of the rover body.
2. Make a set of wheels in a shape of your choice that is identical out of cardboard. Using your ruler to help, poke axle holes into the center of each wheel. Attach the wheels by first pushing the pencil through the set of holes in your rover body. Slide the wheels onto the pencil on either side and secure them to the pencil with tape.
3. Tape your straw to the bottom of your rover body at the opposite end. Slide the 2 hard candies on either side as your second set of wheels. Bend the ends of the straw and tape them to stop the candies from falling off.
4. Make a chain with the 2 rubber bands. Loop the chain around the pencil. Cut two slits in the rover body on the other side by the candies and loop the other link of your rubber band chain through the slits to keep it in place.
5. Wind up your pencil wheels as much as you can, set your rover on the ground, and release! Happy racing!

Rocket Power

How high can your rocket go?

Check out the HTH Physics Book to learn how and why rockets work!

Materials: eye protection, paper or construction paper, markers, tape, scissors, water, fizzy cold medicine tablets, and a 35mm film canister with the lid that fits inside it.

Procedures:

1. Using your film canister as your "clothes" model, design your rocket using paper, colored construction paper, markers, scissors, and tape. You will tape your rocket to the canister, but make sure the lid side is at the bottom of your rocket.
2. Next, you get to be a scientist and determine how much of the cold medicine tablet and how much water you will add into the film canister. Typically 1/2 a tablet and 1/3 a canister of water works nicely. But, why not try other combinations?
3. When you are ready to test your rocket, put on your protective eyewear. Add your water to your canister, drop in your medicine tablet, and quickly shut the lid tight. Set your rocket down on your launch pad and start your countdown!

Engineering Challenge!

Can you keep "astronauts" safe with limited starting materials?

1. Using only 5 sheets of paper and 12 inches of tape, design a spacecraft that could safely land with an astronaut cargo, aka boiled egg.
2. Once you have your prototype complete, place a boiled egg inside and release the structure from different heights checking the damage to the egg each time. You may want to have a few eggs on hand.

Fluid Fluctuations

Microgravity affects the fluid balance in your body. On Earth, gravity is constantly pulling blood towards your legs and feet but in space, fluid balances out equally around your body. This can lead to astronauts having puffy faces and skinnier legs! You can mimic this fluid shift with this simple experiment.

1. While standing for 10 minutes have a partner place pieces of tape at three different locations on your leg: your ankle, your calf, and just above the knee. Near the end of the 10 minutes have your partner measure the circumference (all around the leg) at each spot with a tape measure and record the data.

2. Once that 10 minutes is complete, lie down with your legs propped up against a wall for 10 more minutes. Immediately after, have your partner measure those same three locations again as you stay in that position. Compare your data. What happened? Did you experience any other body symptoms?

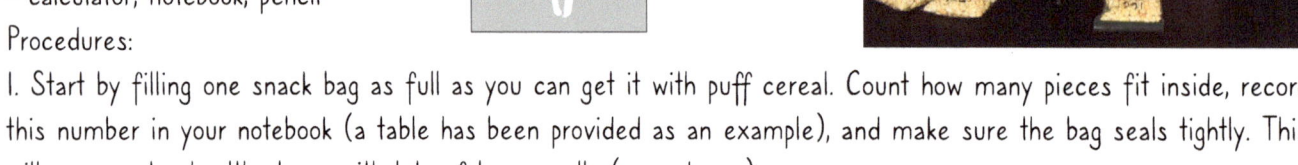

For more on how the lack of gravity affects humans, Go to page 31…

Bone Density

Turns out microgravity causes bone cell loss in astronauts, like osteoporosis here on earth. This can affect their health once they return to Earth.

Materials:
- corn puff cereal
- 3 plastic snack bags
- a very heavy and large book
- calculator, notebook, pencil

Procedures:

1. Start by filling one snack bag as full as you can get it with puff cereal. Count how many pieces fit inside, record this number in your notebook (a table has been provided as an example), and make sure the bag seals tightly. This will represent a healthy bone with lots of bones cells (very dense).

2. Next, you will prepare the second bag to represent 25% bone loss and the third bag to represent 50% boss loss. To do this, take your total number for bag 1, and multiply it by 0.75 (75% remaining bone cells) for bag number 2. This is the number of cereal you will count and place inside sandwich bag 2. Multiply the total from bag 1 by 0.5 (50% remaining bones cells) and count that number to be placed in bag 3. Record both numbers in your notebook.

3. Now that you have the 3 bags of different bone densities, you are going to subject them to a traumatic event, aka book drop. With each bag placed on a flat surface, drop the very heavy book from the same height onto each bag.

4. Count how many pieces of cereal remained whole after the book drop in each bag and record this in your notebook.

5. Calculate the number of bone cells lost by subtracting each bag's remaining cereal from each bag's beginning number of cereal. Take these answers and divide by each bag's original number to get the percentage of bone loss in each case. Was there a big difference in bone loss between the healthy bone (bag 1) and one that was very unhealthy (bag 3)?

Beginning Cereal Count	Cereal Remaining	Cereal Lost (Beginning - Remaining)	Percent Bone Loss ((Lost / Beginning) x 100)
500	350	500-350 = 150	(150 / 500) x 100 = 30%

APPENDICES

Key Vocabulary

Asteroid - small rocky objects (smaller than planets) that orbit the sun.

Axis - an imaginary line around which an object rotates.

Comet - a combination of frozen gas, rock, and dust that heat up as they get closer to the sun in their orbit.

Fusion - the process by which stars like our sun get power.

Galaxy - a huge group of stars, gas, and dust that are held together by gravity.

Nebula - a large cloud of dust and gas.

Newton's Law of Universal Gravitation - every object in the universe is attracted to other objects in the universe.

Neutron Star - the remains of a high mass star.

Orbit - a curved path an object takes around a larger object due to the attractive force of gravity.

Pulsar - a spinning neutron star.

Radiation - energy that comes from electromagnetic waves.

Rover - a man-made planetary surface exploration robot.

Satellite - a man-made object, moon, or planet that orbits another planet or star.

Space Probe - a spacecraft with no crew designed to collect scientific data for specific missions.

Weather - the uneven heating of the earth's air and water due to the earth's tilt and distance from the sun.

Books in the Series!

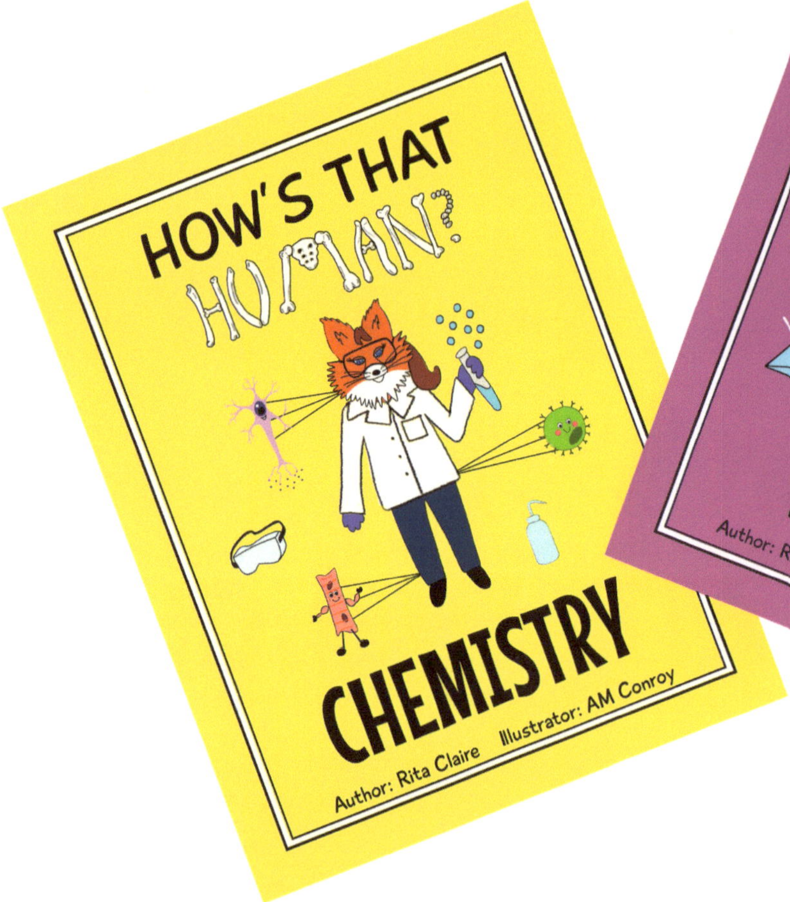

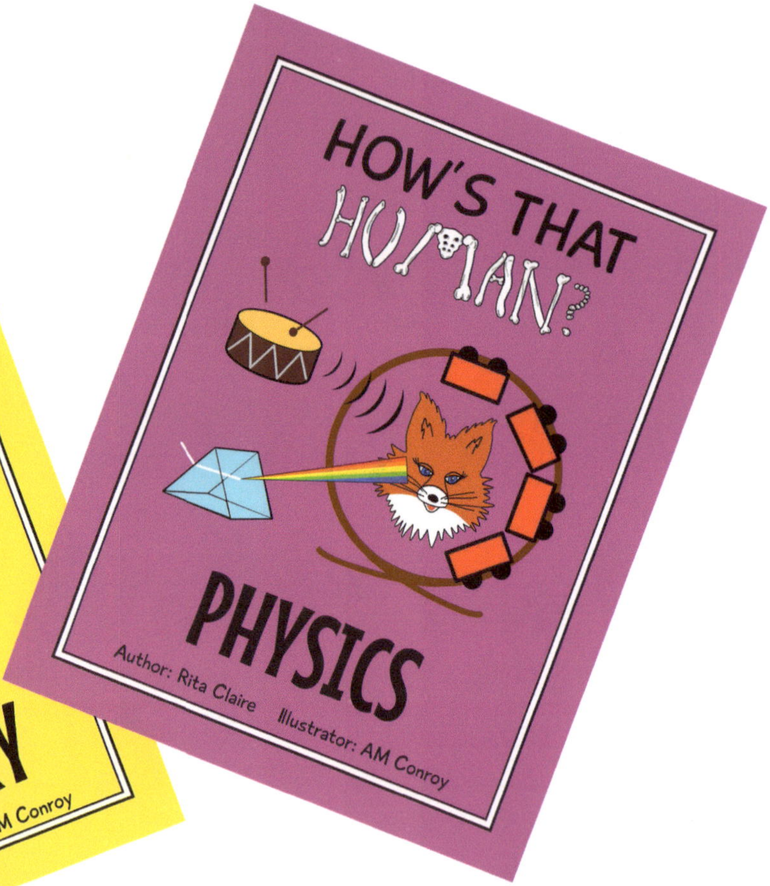

Website Tips!

https://www.howsthathuman.com/

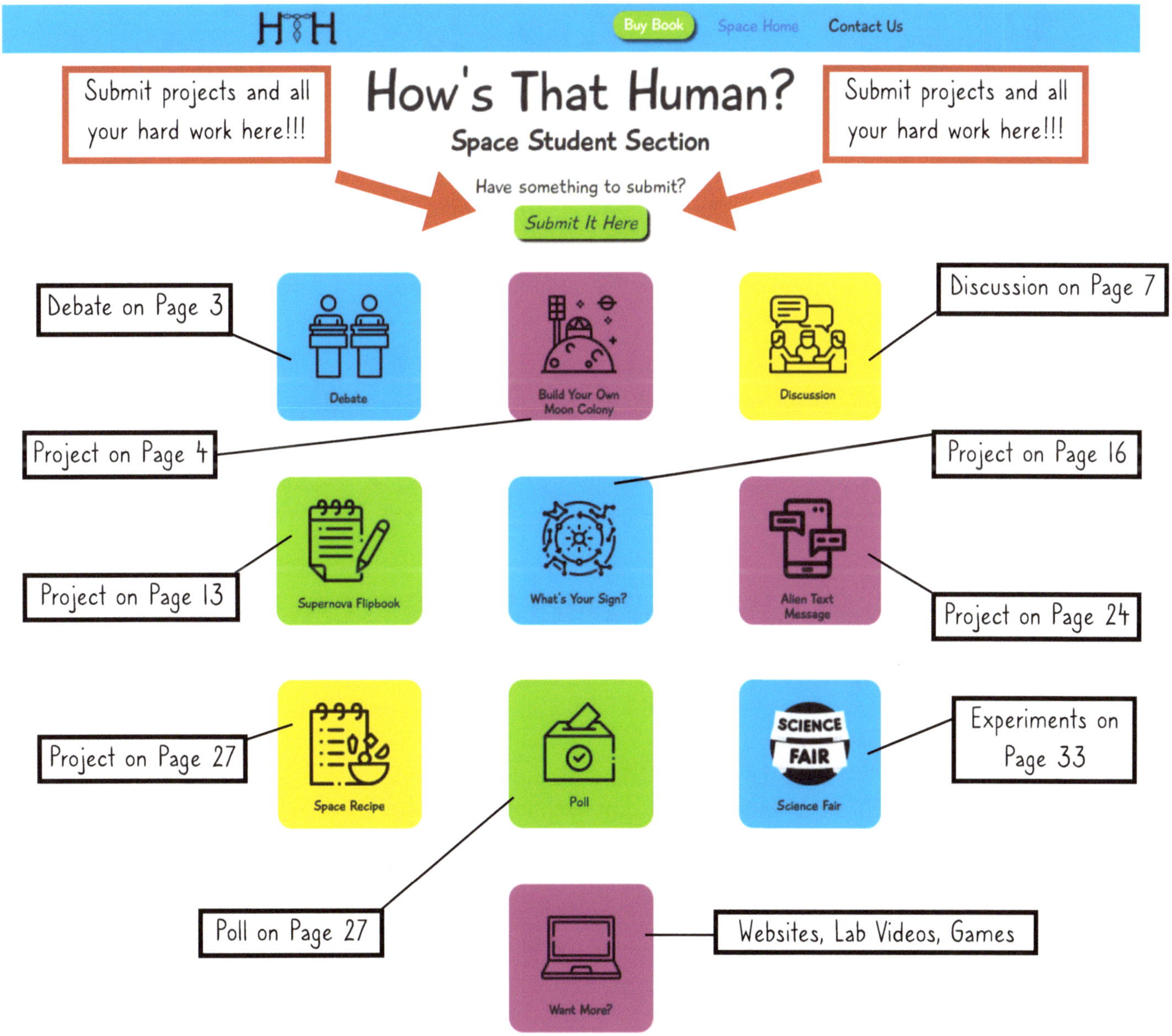

Start by clicking on the blue Space box!

Submit projects and all your hard work here!!!

Submit projects and all your hard work here!!!

Debate on Page 3

Project on Page 4

Discussion on Page 7

Project on Page 13

Project on Page 16

Project on Page 24

Project on Page 27

Experiments on Page 33

Poll on Page 27

Websites, Lab Videos, Games

42

Scientist Spotlight

Author Rita Claire is a former molecular virologist and geneticist who worked on plant virus research involving cereal crops and human genetics projects. She misses research, but has thoroughly enjoyed teaching science K-12 and beyond. Ms.C, as she's affectionately called by her students, is passionate about changing the way we educate and has always dreamed of being an author. As a recent homeschool mom, she wanted to create a more relatable educational tool for other homeschool or distance learning families, as well as elementary school districts. Welcome to How's That Human?

Illustrator AM Conroy has a Ph.D. in biology from UCLA and has worked in marine biology, animal physiology, and biomechanics. Her love for animals inspired her to study the movements of amazing creatures such as swimming pufferfishes, running tigers, and hopping kangaroos. Her mom, an illustrator as well, encouraged her interest in drawing as a child, with favorite subjects being snails, fishes, and unicorns. She enjoys illustrating for How's That Human? and hopes that her drawings help excite learners of all ages!

Acknowledgments

Amina Filkins from Pexels
Lightfieldstudiosprod | Dreamstime.com

Special Thanks: It is with much gratitude and humility that I want to thank NASA and all of its affiliates/partners worldwide for the images and information in this text. Without your continued curiosity and perseverance in space exploration the human race would not be where it is today!